Discoverin[g]

Shellfish

by Melvin and Gilda Berger

SCHOLASTIC INC.

No part of this publication may be reproduced,
stored in a retrieval system, or transmitted in any form
or by any means, electronic, mechanical, photocopying,
recording, or otherwise, without written permission of the
publisher. For information regarding permission,
write to Scholastic Inc.,
Attention: Permissions Department, 557 Broadway,
New York, NY 10012.

ISBN 978-0-545-35172-0

Copyright © 2013 by Melvin & Gilda Berger

All rights reserved. Published by Scholastic Inc.
SCHOLASTIC and associated logos are trademarks
and/or registered trademarks of Scholastic Inc.

12 11 10 9 8 7 6 5   16 17 18/0

Printed in the U.S.A.   40
First printing, May 2013

Photo Credits: Photo Research: Alan Gottlieb

Cover: © Jens Kuhfs/Getty Images; Back cover: © William H. Mullins/Science Source/Photo Researchers; Title page: © Visuals Unlimited, Inc./Gerald & Buff Corsi/Getty Images; page 3: © Marevision/Getty Images: page 4: © Martin Shields/Science Source/Photo Researchers: page 5: © blickwinkel/Alamy; page 6: © Lew Robertson/Getty Images; page 7: © Solvin Zankl/Nature Picture Library; page 8: © Amar and Isabelle Guillen-Guillen Photography/Alamy (RF); page 9: © Richard Waters/Shutterstock; page 10: © iStockphoto; page 11: © Taesam Do/Getty Images; page 12: © Philip Plisson/Nature Picture Library; page 13: © Andrew J. Martinez/Seapics.com; page 14: © Andrew J. Martinez/Science Source; page 15: © Tetra Images/Getty Images (RF); page 16: © Bob Elsdale/Getty Images

Shellfish are sea animals with shells.

3

All shellfish have hard outer shells.

Can you find the shellfish?

The shells keep them safe.

Some kinds of shellfish are like lobsters, crabs, and shrimp.

How many feelers do you see?

They have feelers on their head.

Are both claws the same size?

The front legs have claws.

The back legs help the animals walk or swim.

Other shellfish are like clams and oysters.

Can you see inside the shells?

Their shells can open and close.

Their soft bodies are inside the hard shells.

Does the foot look like your foot?

Some have a foot that digs in mud or sand.

Is this shellfish walking or swimming?

Many shellfish live at the bottom of the sea.

People like to collect shells.

15

Ask Yourself

1. What do all shellfish have?
2. Are all shellfish alike?
3. Can you name a shellfish with claws?
4. Can you name a shellfish with two shells?
5. Where do many shellfish live?

You can find the answers in this book.